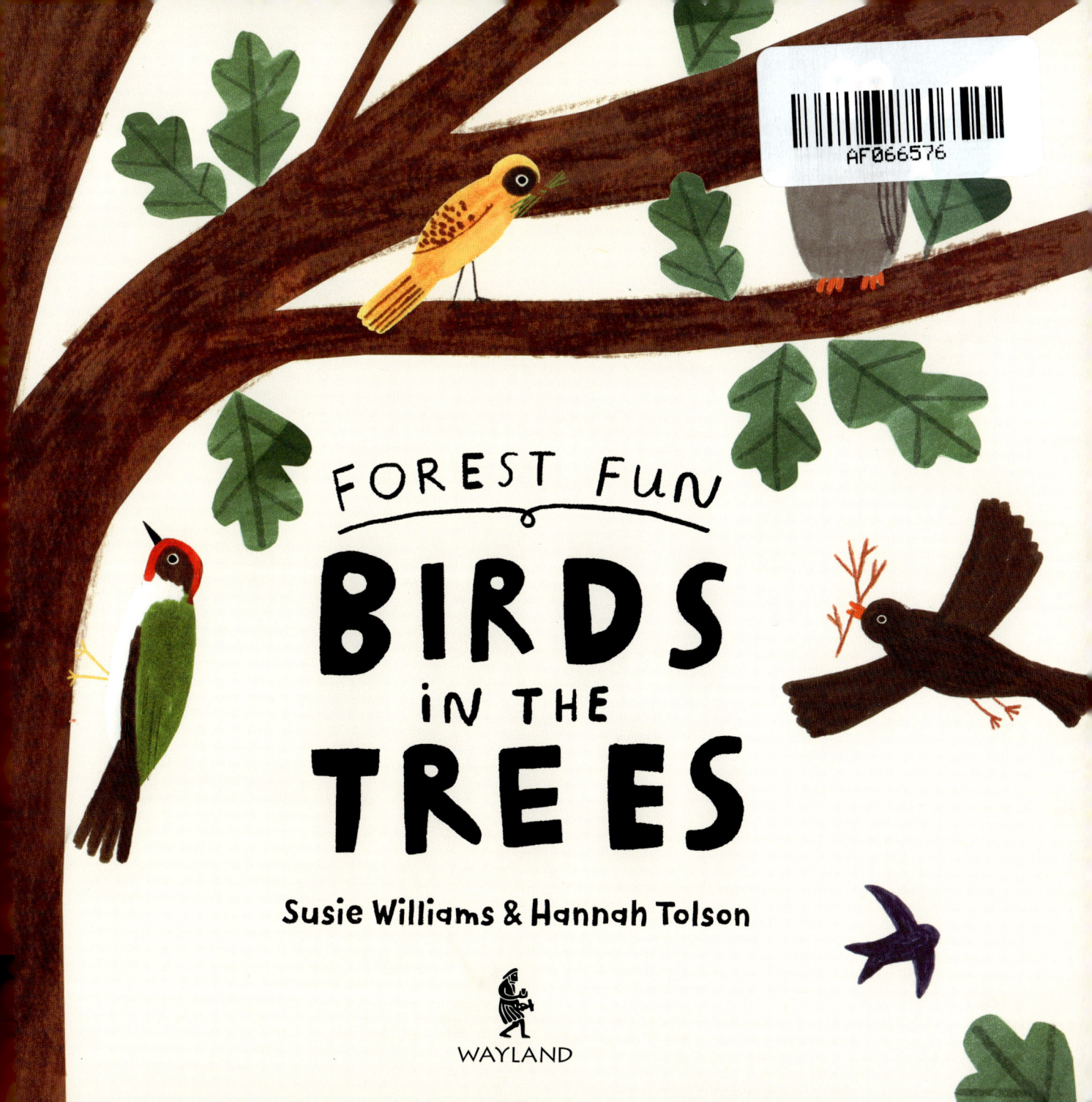

FOREST FUN
BIRDS IN THE TREES

Susie Williams & Hannah Tolson

WAYLAND

First published in Great Britain in 2024
by Wayland

Copyright © Hodder and Stoughton, 2024

All rights reserved

Editor: Victoria Brooker
Designer: Lisa Peacock

ISBN: 978 1 5263 2343 9 (hbk)
ISBN: 978 1 5263 2344 6 (pbk)

Printed and bound in China

Wayland, an imprint of
Hachette Children's Group
Part of Hodder and Stoughton
Carmelite House
50 Victoria Embankment
London EC4Y 0DZ
An Hachette UK Company
www.hachette.co.uk
www.hachettechildrens.co.uk

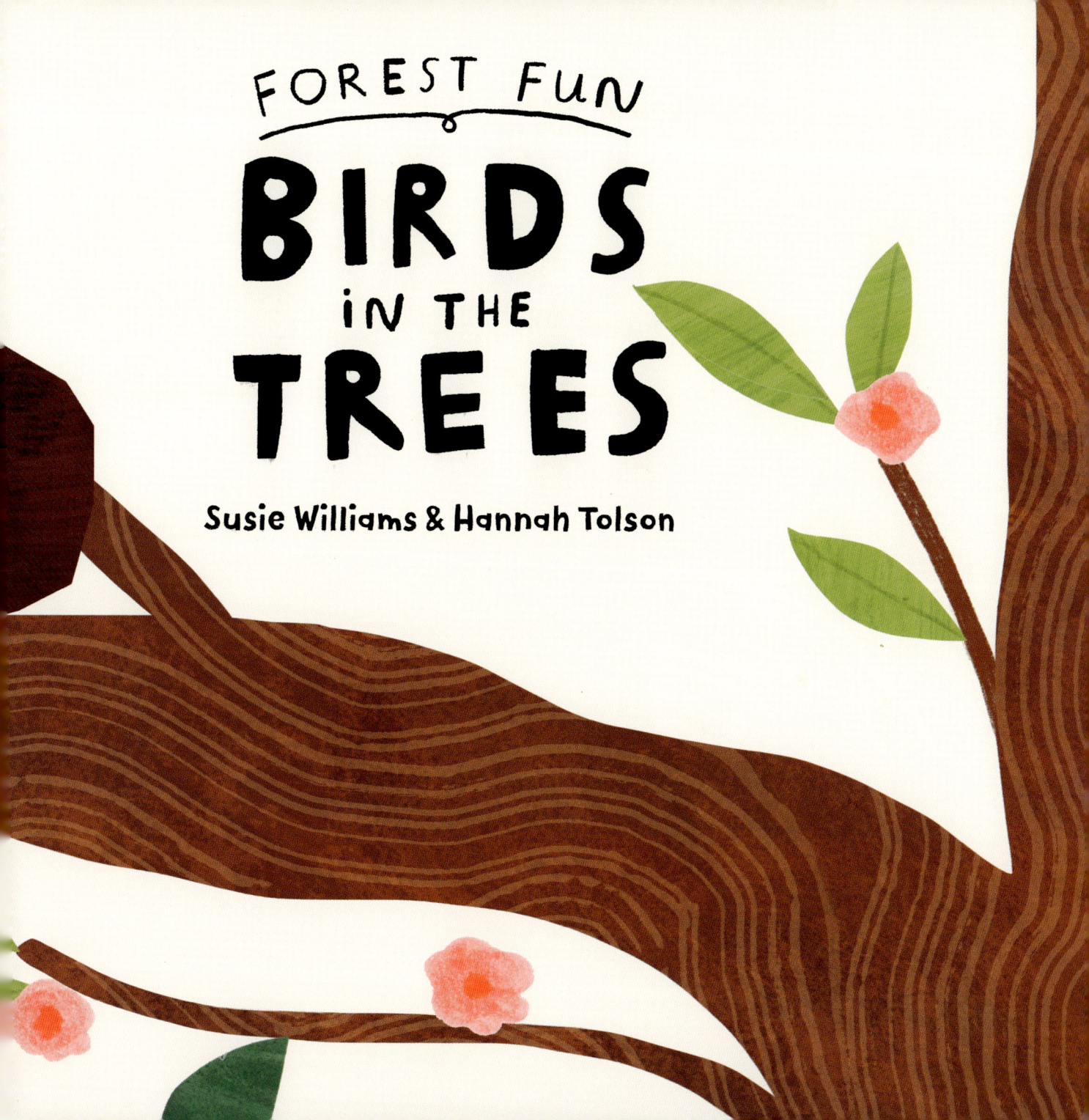

A forest is a place that has lots and lots of trees.

Forests are home for many kinds of animals and plants.

You can see birds soaring above the trees, perched in the treetops or pecking for worms on the forest floor.

There are different layers in a forest, such as the forest floor, the undergrowth and the canopy, also called the tree layer.

Most birds live in the tree layer. Here, they are protected from other creatures that might eat them, such as bears and wolves.

Hundreds of birds live in a forest.

Birds communicate by singing, chirping and squawking to each other.

They use their calls to warn other birds about danger, scare away predators or attract a mate.

Some birds, such as nuthatches and woodpeckers, scoot up and down the tree trunks.

They live in very small holes in tree trunks.

At the foot of trees, some birds scurry around the base.

They don't go up high in the trees but stay low in the undergrowth.

Their brown feathers camouflage them so they can hide from predators.

Many birds roost at night, resting in the trees to keep safe.

But not all birds are asleep.

Birds lay eggs in nests to protect them from predators and bad weather.

Birds stay on their nests, keeping the eggs warm so that the babies will hatch safely.

When the baby birds are born, they are blind and helpless.

Their parents collect thousands of insects and grubs for them to eat.

In summer, some birds arrive in the forest from other countries.

They come to lay their eggs in the warm weather while there are lots of insects to feed their babies.

They will have travelled for thousands of miles and will leave again when the weather gets colder.

When the leaves start to change colour in autumn, there are many fruits on the trees.

Finches like to feast on the juicy berries.

They try to feed up for the winter when there will be less food around for them.

Seed-eating birds, such as crossbills, help the forest by opening the cones and letting seeds fall to the ground.

The seeds grow into new trees.

In winter, feathers are very important for keeping birds warm.

By fluffing up their feathers, they trap pockets of warm air around their bodies.

Small birds crowd together to share body heat.

Birds like nuthatches and woodpeckers keep warm in tree hollows.

See if you can spot any forest birds in winter.

Make a bird feeder

During winter, some birds struggle to find food. Make your own bird feeder to hang in your garden, school yard or at a wildlife centre.

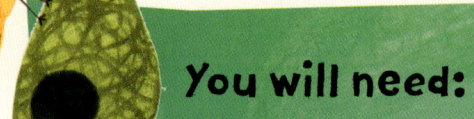

You will need:
- Yogurt pot
- Lolly stick
- String
- Some oats, raisins, breadcrumbs, unsalted peanuts, stale cake crumbs or sunflower seeds
- Lard

1. Ask an adult to help you use a pencil to make a hole in the bottom of a yogurt pot.

2. Tie a length of string to a lollipop stick and thread the other end through the yogurt pot hole.

3. Mix together some oats, raisins, breadcrumbs, unsalted peanuts, stale cake crumbs or sunflower seeds. You can choose what to include.

4. Add an equal amount of chopped squares of lard to the mixture. Squeeze it with your fingers to make a ball of lard and nut mix.

5. Spoon the lard mixture into the yogurt pot. Leave it in the fridge for an hour to harden.

Once it is set, slide the yogurt pot off and hang your bird feeder from a tree branch. Move away and then watch quietly as the birds come and feast!
How many different birds can you spot?

More wonderful facts about birds

Every year, birds moult (lose their feathers) and grow a whole new set to replace them.

Flapping wings takes a lot of energy. Small birds flap their wings in short, fast burst. Large birds flap slowly and glide to save energy.

Birds' eyes take up more space in the head than the brain does.

Did you know that:

- birds have no teeth? They have to swallow all their food without chewing.

- some birds are very clever and can learn from others!

Glossary

bark – the outside part of a tree

camouflage – a way of hiding by looking like the surrounding environment

mate – one of a pair of animals that have babies together

perch – a place, such as a branch, where birds rest

predator – an animal that hunts other animals for food

prey – an animal hunted and eaten by another animal

roost – to sit and rest on a perch

trunk – the main stem of a tree

Index

autumn 22

bark 11

camouflage 13
canopy 6
communicate 8

eggs 17, 18, 20

feathers 13, 16, 26
food 23, 28, 29
forest floor 5, 6

insects 11, 19, 20

owls 15

nests 16, 17, 18

predators 9, 13

roost 14

seeds 24, 25, 28
spring 16
summer 20

trees 4, 5, 11, 12, 14, 15, 16, 25
tree hollows 15, 27
tree trunks 10
treetops 5

undergrowth 6, 12

winter 23, 26, 27, 28
worms 5